Praise for *My Place in Space*

'A book to pore over, again and again.'
Kirkus Reviews

'One for the parents and children to learn from as they chuckle.'
Booklist

'*My Place in Space* will introduce youngsters to astronomy, add humor to primary-grade science lessons, or just keep the kids occupied for a time trying to see all that's going on in the detailed drawings.'
Starred Review, *School Library Journal*

'The collaboration of scientists and illustrators on this book will give pleasure to readers with interests that range from the macrocosmic to the microcomic.'
Starred Review, *Horn Book*

'*My Place in Space* is a fun way to introduce children to the wonders of astronomy and awaken their curiosity ... With vibrant illustrations and accurate text even adults will enjoy it and be left wanting to know more.'
***** Daniel C. McGuire, Amazon

Honour Book: Children's Book Council of Australia, Picture Book of the Year
Pick of the Lists, *American Bookseller*, USA
Bulletin Blue Ribbons, *Bulletin of the Center for Children's Books*, USA
Notable Children's Trade Book, Field of Social Studies NCSS–CBC, USA
Notable Trade Book, Language Arts, NCTE, USA

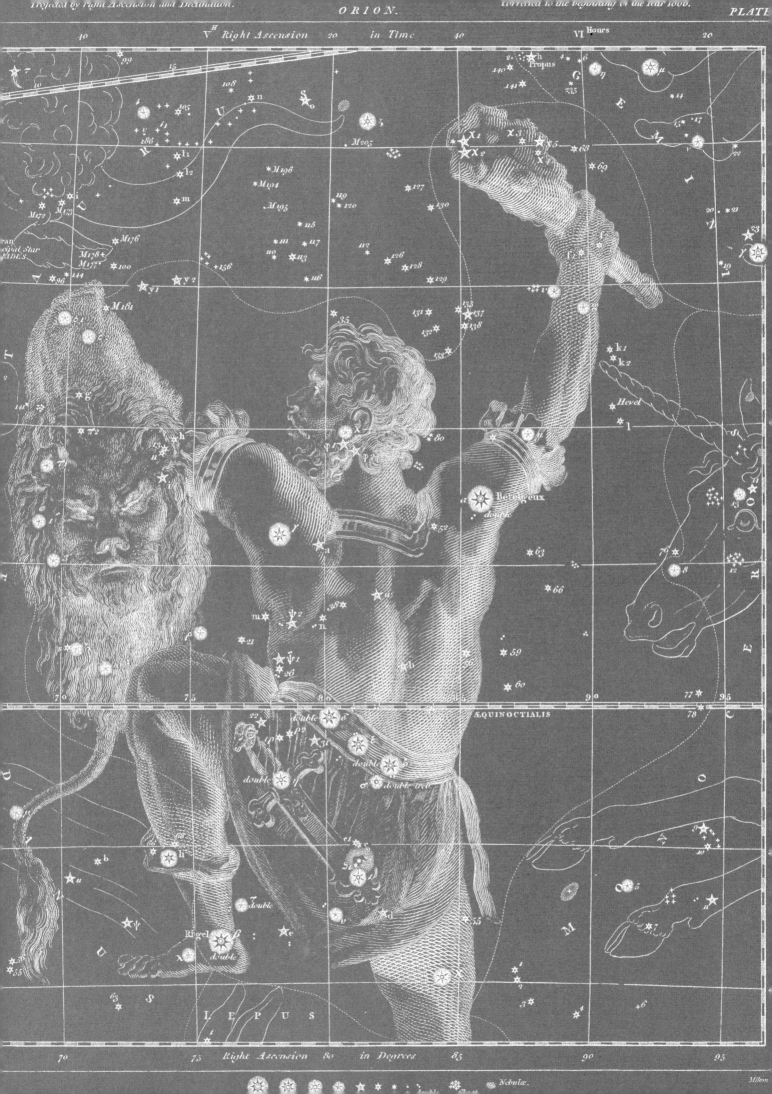

First published by The Five Mile Press in 1988
This revised and updated edition published in 2008

Allen & Unwin
83 Alexander Street
Crows Nest NSW 2065
Australia
Phone: (61 2) 8425 0100
Fax: (61 2) 9906 2218
Email: info@allenandunwin.com
Web: www.allenandunwin.com

National Library of Australia
Cataloguing-in-Publication entry:

Hirst, Robin A.
My place in space.
Rev. ed.
For children.
ISBN 978 174175 404 9 (pbk.).

1. Home – Juvenile literature. 2. Cosmological distances – Juvenile literature.
3. Southern sky (Astronomy) – Juvenile literature. 4. Expanding universe – Juvenile
literature. 5. Solar system – Juvenile literature. I. Hirst, Sally. II. Levine, Joe.
III. Harvey, Roland, 1945– . IV. Title.

523.1

Cover and design by Ruth Grüner
Orion & Ursa Major and Ursa Minor Constellation reproductions (endpapers)
from Longman, Hurst, Rees & Orme publication, 1808
Typeset in Harvey created by Sandra Nobes from Roland Harvey's handwriting

Printed by Everbest Printing Co., China

3 5 7 9 10 8 6 4 2

MY Place IN Space

THIS BOOK BELONGS TO . . . .

WHO LIVES AT . . . .

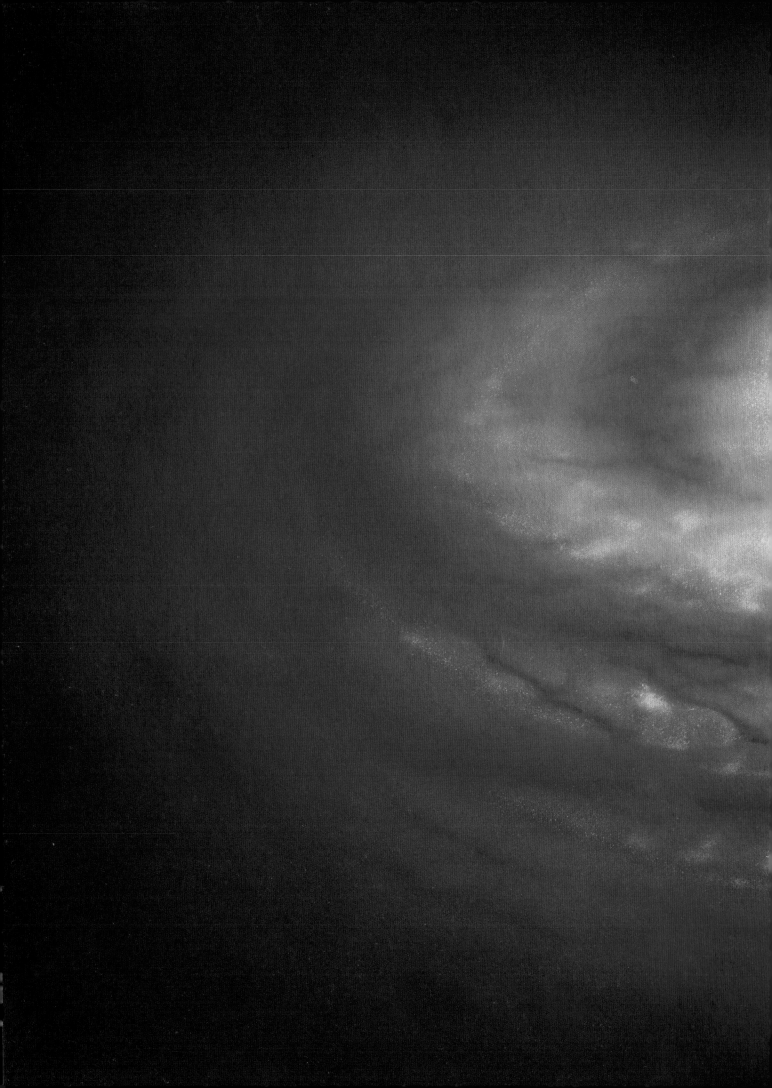

ROBIN & SALLY HIRST

My Place in Space

Illustrated by
Roland Harvey
& Joe Levine

ALLEN&UNWIN

Not so long ago, Henry and Rosie Wilson were waiting for a bus. The bus roared around a corner and, with a sudden swerve towards the kerb and a hiss from the brakes, it stopped in front of them.

The door swung back. The tall and rather scruffy driver peered down at Rosie and Henry.

'Who are you?' asked the driver. 'And where would you like to go?'

   'I'm Henry Wilson and this is my sister, Rosie,' said Henry. 'We'd like a ride home, please.'

'Home?' asked the driver loudly, imitating Henry's voice and raising his eyebrows.

Henry felt his face go red.

'Well,' said the driver, leaning forwards, 'that depends on whether you can remember where you live!'

He laughed and winked at two women on the front seat of the bus. They giggled.

'Do you know where you live, Henry Wilson?' the driver's voice boomed.

The giggling women stared at Henry, waiting for his reply.

Henry gritted his teeth and took a deep breath.
  'Yes, I do know where we live. We're at number
12 Main Road, Gumbridge. Our house is on the north side of
Main Road. The town of Gumbridge is just over the river from
here, about twenty kilometres south of the Dividing Range.'

 The driver's grin faded slightly.
Henry took a deep breath and continued.
'Gumbridge is a typical country town in Australia.'

'Australia is in the Southern Hemisphere of the planet Earth,' continued Henry.

   The women in the front seat had stopped giggling and their mouths were gaping rather foolishly. So was the bus driver's.

Rosie squeezed Henry's hand. Gaining courage, Henry took another deep breath and went on.

'The planet Earth is one of the eight major planets which circle the star we call the Sun. Earth is the third planet from the Sun, 150 million kilometres away from it.'

'It takes eight minutes for the Sun's light to travel from the Sun to Earth,' said Henry. 'That may seem a long way, but it's 4,500 million kilometres to Neptune, the eighth planet from the Sun. Sunlight takes over four hours to reach Neptune.'

'Shall I tell him about Pluto?' asked Rosie. Henry nodded.
    Rosie turned to the driver. 'In case you were wondering why Henry didn't mention Pluto,' she said, 'it is now called a dwarf planet.'
    The driver said nothing, even though his jaw was moving up and down.
    'We call the Sun and planets our Solar System,' said Henry.

'Our Solar System is in the middle of a group of stars we call the Solar Neighbourhood. Our Sun's neighbours are stars of all colours and sizes. You can see them in the dark skies of Gumbridge on any clear night,' said Henry. 'The nearest star to the Sun that we can see is called Alpha Centauri. Even though it's the closest, it takes over four years for its starlight to reach our Sun.'

'Other stars, like the big red star Antares, are right out to the edge of the Solar Neighbourhood, about 600 light years away — that means that it takes 600 years for the light to reach us,' said Rosie. She smiled at Henry and took another deep breath.

'The Solar Neighbourhood is just a small part of the Orion Arm,' said
Rosie. 'Stars aren't spread out evenly in space — they hang around in
groups. Our Sun is one star in a group of millions of stars that make
a giant curved shape in space. The curve is called the Orion Arm.
In among the stars are huge clouds of dust and glowing gases.'

'You can see the dust clouds from Gumbridge, too,' said Henry.
'They look like dark patches in the very starry parts of the sky.'
The driver checked his watch. He looked as though he was about
to say something, but Henry quickly continued.

'The Orion Arm is just one of the arms of a huge group of about three hundred thousand million stars. This whole group is called the Milky Way Galaxy.'

'If you think it's a long way to the nearest star, that's nothing — it takes light one hundred thousand years to cross from one side of the Milky Way Galaxy to the other,' said Rosie.

 'The Milky Way Galaxy is called a spiral galaxy because of the way its arms make a spiral shape.' Henry and Rosie demonstrated the spiral arms.

 'Of course, the Milky Way isn't the only galaxy of stars in space. There are thirty galaxies in our Local Group of Galaxies. Ours is the second largest. The largest is called the Andromeda Galaxy. It's a spiral galaxy too,' said Henry, making more spiral shapes with his arms.

'Light from the stars in the Andromeda Galaxy takes over two million years to reach our galaxy. Most of space is just that — space,' added Rosie. The bus driver looked very puzzled indeed.

The two women on the front seat were staring at the driver, whose cheeks were turning slightly red.
    'Our Local Group of Galaxies is just part of a huge group of galaxies called the Virgo Supercluster,' said Rosie.

'It takes light a few million years to cross our Local Group of Galaxies, but it can take a few hundred million years to cross from one side of a Supercluster to another.'

The driver's face was now a bright shade of red
and he was shaking his head from side to side.

With one more deep breath and a quick smile, Henry waved both arms to make a large circle and said triumphantly, 'And the whole Universe is filled with Superclusters of Galaxies!'

Rosie stepped in front of Henry. She put her
hands on her hips and pushed out her chest.
'So we do know where we live,' she said.

'We live at 12 Main Road,
Gumbridge, Australia,
Southern Hemisphere . . .

. . . Earth . . .

. . . Solar System . . .

. . . Solar Neighbourhood . . .

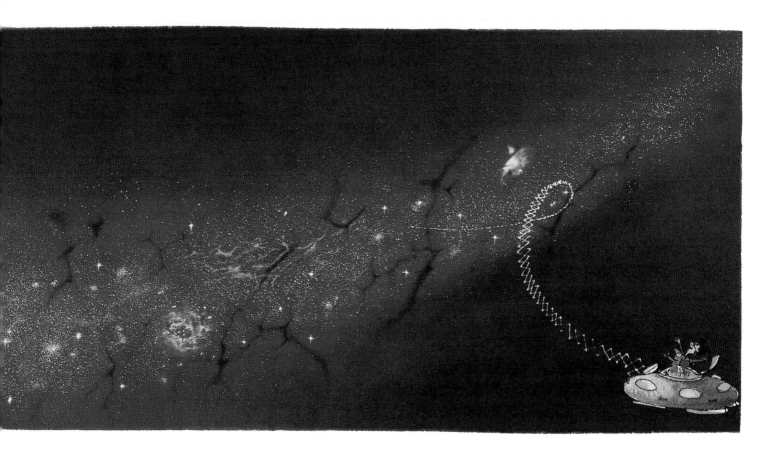

. . . Orion Arm . . .

. . . Milky Way Galaxy . . .

. . . Local Group of Galaxies . . .

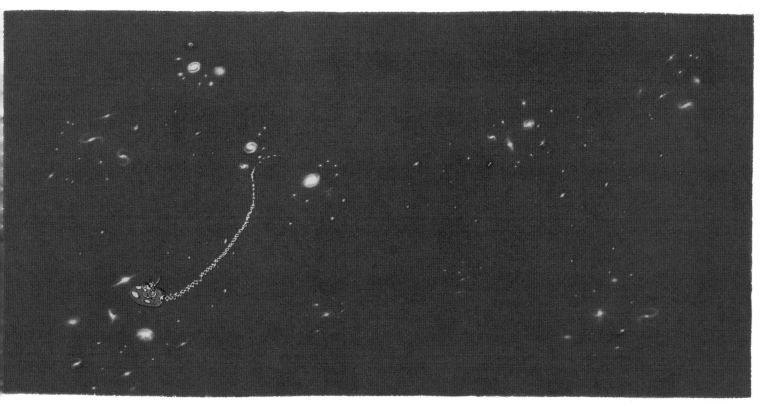

... Virgo Supercluster ...

... the Universe.'

'May we have our tickets, please?' asked Henry.

Understanding the Universe

Understanding space can be tricky. Henry and Rosie show us how to think about the enormous size of the Universe by beginning with the Solar System and moving out to the Superclusters, so it's easier for us to understand the immensity of space, and our place in it.

The distances in outer space are huge. The distance from Earth to the moon is only 384,400 kilometres. From Earth to the Sun is 150 million kilometres. That's 3,500 times around Earth. The nearest star is Alpha Centauri and it's more than 40,000,000,000,000 kilometres away. So we use millions of kilometres to describe the incredible distances to our planets, and light years for things even further away.

A light year sounds like a measurement of time, but it isn't. It's the distance light travels in a certain time. Light travels 300,000 kilometres each second, so a light second is a distance of 300,000 kilometres. (That's seven times around Earth.) Light travels the distance from Earth to the Sun in eight minutes, so that distance is eight light minutes. A light year is the distance light travels in one year, which is about 9,500,000,000,000 kilometres. Our nearest spiral galaxy, Andromeda, is over 2 million light years away!

Building a model of the Solar System is a great way to understand space. The Sun is much bigger than Earth, and they are 150 million kilometres apart. So, if the Sun were the size of a tennis ball, Earth would be the size of a pin head about seven metres down the hallway. The nearest visible star, Alpha Centauri, would be another tennis ball 2,000 kilometres away! Pluto would be a grain of sand down the street in a neighbour's backyard.

Cities are not the best places to view the stars. We struggle to see them against the city lights. From Henry and Rosie's home in the country, they can see a few thousand stars, and even the Milky Way which looks like a whitish glow in a band across the sky. If you look carefully at the brightest stars you might notice their different colours. Some seem to glow orange and some seem blue. Next time you are outside at night-time, take a look!

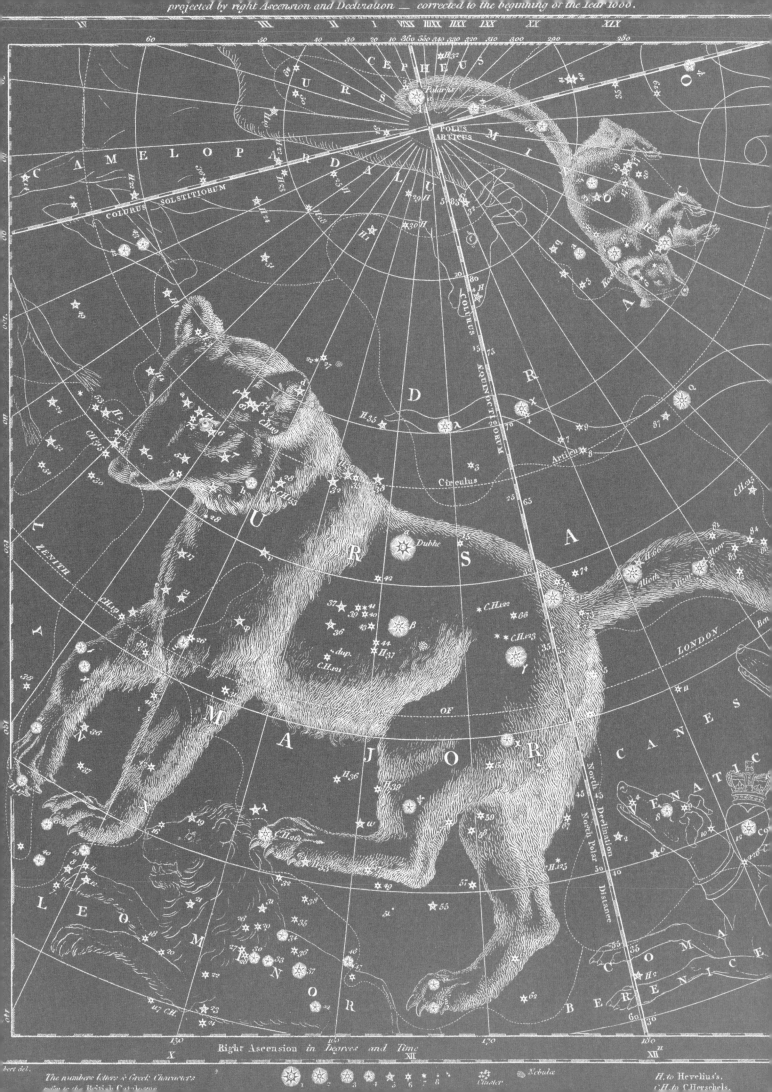

CEPHEUS

CAMELOPARDALUS

COLURUS SOLSTITIORUM

Polaris

POLUS ARCTICUS

D R A C O

Dubhe

U R S A

Alioth

Alcor

Mizar

LONDON

CANES

VENATICI

North Declination

North Polar Distance

U R S A M A J O R

ZENITH

Circulus

Arcticus

AQUINOCTIORUM

COLURUS

L E O

L E O M I N O R

C O M A

B E R E N I C E

The numbers Stars & Greek Characters refer to the British Catalogue.

Cluster   Nebula

H. to Hevelius's.
C.H. to C.Herschels.

ORION.

Robin Hirst

Robin Hirst migrated from England to Australia when he was six. He has no recollection of seeing any stars from the industrial city of Leeds in England, but from the outskirts of Ballarat he could see spectacular star-studded black skies and the glowing band of the Milky Way. At the University of Melbourne Robin studied and taught astronomy, and he has a doctorate in Infrared Astronomy. In 1981 he was appointed Director of the Melbourne Planetarium with Museum Victoria, and wrote scripts and produced planetarium shows. Robin is currently Director, Collections Research and Exhibitions for Museum Victoria, Australia's largest museum organisation.

Sally Hirst

Sally Hirst was born in Wales and grew up in Singapore and the United Kingdom. She graduated from the University of London and came to Australia in 1975. She planned to teach in Australia for eighteen months, and then she forgot to go home. Sally met Robin at the Museum and they began writing for children. They also played key roles in the development of Scienceworks, the popular science and technology centre, and Sally was a principal of the Museum Education Service. Sally is a director of a consultancy company and uses her creative talents to solve problems and develop business ideas and marketing plans for places such as museums, galleries, libraries and parks and gardens around Australia.

Roland Harvey

Roland Harvey is one of Australia's best-loved illustrators. In 1978 he established Roland Harvey Studios where he produced greeting cards, posters and stationery with a distinctly Australian flavour. Ventures into publishing quickly followed. His books include *Burke and Wills*, *The First Fleet* and *Sick As* (with Gael Jennings), the bestselling *At the Beach*, *In the Bush* and *In the City* and his illustrations are featured in the *Bonnie & Sam* series – *The Horses and Ponies of Currawong Creek*, in collaboration with Alison Lester.

Joe Levine

Joe Levine is a prolific freelance illustrator, based in Melbourne. Joe was born in London and left school at sixteen to work in an art studio and attend life classes at St Martin's School of Art in the evenings. At 21 he set off on a hitchhiking adventure and six months later arrived in Melbourne in August 1960 where he took up freelance illustration work. He has a Bachelor of Arts (Eng. Lit), is currently completing a master's degree in Renaissance Studies and he has a lively interest in literature, classical music, jazz, judo and karate.